Christmas
Math
For First Grade

This book belongs to

_____

Liv & Blue Publishing, LLC

Table of Contents

 Operations & Algebraic Thinking

 Measurement & Data

 Number & Operations in Base Ten

 Geometry

4 Stocking Stuffers

12 Give the Mitten

14 Glitters Gold

16 Let It Snow

24 Most Wonderful Time

26 Santa's Workshop

28 Merry and Bright

36 Gravy Trains

38 Stocking Up or Down

40 Boughs of Holly

44 Red-Nosed Reindeer

46 All Shapes and Sizes

48 Shape Your Tree

6 Three's a Charm	8 Present and Correct	10 Snowglobe Trotters
18 Dapper Little Guy	20 Oh Christmas Tree	22 For Good Measure
30 Clans of Gingerbread	32 Frosty Together Again	34 Ten's the Season
42 Candy Cane Lane		
50 Season's Greetings		

Student Reflection

After completing an activity, come back to this page to reflect on how it made you feel. In the tile's white circle, draw the snowman's face that matches how you felt about the activity.

Great! This was a lot of fun and I learned a lot!

Good. This was interesting and I understood it.

So-so. This was a little hard, but I learned something.

Bad. This was hard and I still don't understand it.

Stocking Stuffers

Read the word problems to figure out which number is hiding inside the stocking, and then write it in.

There were already some ornaments on a tree. Then a child hung 5 more ornaments to make a total of 12 ornaments. How many ornaments were on the tree to begin with?

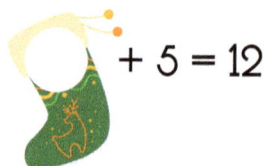

$$\boxed{} + 5 = 12$$

One day Santa received 7 letters from kids around the world. The next day he checked his mailbox and found 7 more letters. How many letters did Santa receive in all?

$$7 + 7 = \boxed{}$$

One Saturday a family baked 16 cookies. Throughout the day the family members ate 11 cookies. How many cookies did they have left at the end of the day?

$$16 - 11 = $$

There were 9 surprises in a boy's stocking one night, but only 5 surprises in the stocking in the morning. How many surprises did the boy sneak out in the middle of the night?

$$9 - = 5$$

Three's a Charm

Read the word problems and write the missing numbers in the top equation. Beneath it, add the first two numbers from the top together, and then add that sum to the third number to get your answer.

A brother and sister decided to bake gingerbread cookies. They baked 4 gingerbread men, 6 gingerbread stars, and 3 gingerbread houses. How many cookies did they bake in all?

$$\_\_ + \_\_ + \_\_ = \underline{?}$$

$$\underline{\hspace{2cm}} + \_\_ = \_\_$$

A class of first graders decorated the school with stockings. They hung 8 snowflake stockings, 4 green stockings, and 2 candy-filled stockings. How many stockings did they hang in all?

$$\_\_ + \_\_ + \_\_ = \underline{?}$$

$$\underline{\hspace{2cm}} + \_\_ = \_\_$$

A family decorated the Christmas tree with ornaments. The uncle hung 5 purple ornaments, the mother hung 7 red ornaments, and the grandpa hung 3 green ornaments. How many total ornaments were on the tree?

_____ + _____ + _____ = ?

_____ + _____ = _____

For Christmas three friends received presents. On friend received 4 yellow presents, one friend received 4 red presents, and one friend received 4 purple presents. How many presents did they receive in all?

_____ + _____ + _____ = ?

_____ + _____ = _____

Present and Correct

If you know the answer to one addition problem, you know the answer to another addition problem in which the numbers are just switched around.

Example:

If $3 + 4 = 7$,

then $4 + 3 = 7$

Now you try! Draw the missing presents and complete the equation.

If $4 + 1 = 5$,

then $1 + 4 =$ _____

If $3 + 5 = 8$,

then $5 + 3 =$ _____

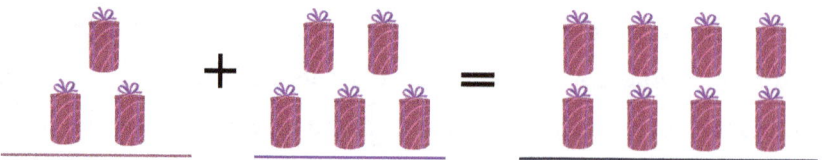

Operations & Algebraic Thinking

When adding three numbers, find out if two of them add up to 10. If so, add them together first, and then add the third number to the 10.

Number pairs that add up to 10:

0 + 10	10 + 0
1 + 9	9 + 1
2 + 8	8 + 2
3 + 7	7 + 3
4 + 6	6 + 4
5 + 5	

 $+$ $= 10$

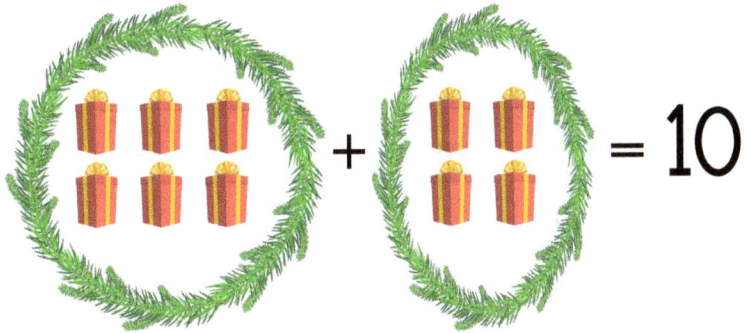

Circle two groups of presents that add up to 10, and then solve the problem.

 $+$ $+$

$$4 + 3 + 7 = \underline{\hspace{1cm}}$$

 $+$ $+$

$$6 + 2 + 8 = \underline{\hspace{1cm}}$$

 Snowglobe Trotters

Solve the subtraction problem by drawing more snowglobes until they add up to the first number in the equation. The number of snowglobes you draw is the answer to the problem.

$7 - 5 = \underline{\hspace{1cm}}$

$4 - 1 = \underline{\hspace{1cm}}$

$6 - 3 = \underline{\hspace{1cm}}$

$7 - 2 = \underline{\hspace{1cm}}$

$8 - 6 =$ ____

$7 - 4 =$ ____

$9 - 7 =$ ____

$8 - 5 =$ ____

Give the Mitten

Count forward on mitten fingers to add numbers.

Example: 4 + 2 = ____

Write the first number on the mitten.

Check as many fingers as the second number.

Count on from the first number to the last checked finger to get your answer.

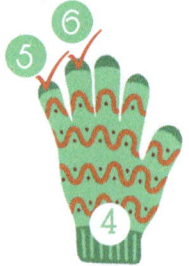

4 + 2 = _6_

Now you try!

4 + 3 = ____

8 + 2 = ____

5 + 5 = ____

1 + 4 = ____

3 + 1 = ____

9 + 0 = ____

CCSS.MATH.CONTENT.1.OA.C.5

Count backward on mitten fingers to subtract numbers.

Example: 6 — 3 = ____

Write the first number on the mitten.

Check as many fingers as the second number.

Count back from the first number to the last checked finger to get your answer.

6 — 3 = __3__

Now you try!

6 — 0 = ____

5 — 1 = ____

8 — 4 = ____

5 — 5 = ____

6 — 2 = ____

7 — 3 = ____

Glitters Gold

Reflect the number order in the equations through the shiny baubles to explore the relationship between + and −.

Example:

$1 + 3 = 4$ $4 - 3 = \underline{1}$

Now you try!

$9 - 5 = 4$ $4 + 5 = \underline{}$

$10 + 7 = 17$ $17 - 7 = \underline{}$

$13 - 8 = 5$ $5 + 8 = \underline{}$

$0 + 6 = 6$ $\underline{} - 6 = 0$

$14 - 5 = 9$ $9 + \underline{} = 14$

$8 + 8 = 16$ $\underline{} - 8 = 8$

$19 - 9 = 10$ $10 + \underline{} = 19$

Let It Snow

Draw the equal sign ═ in the wreath between the snow flurries that add up to the same number of snowflakes on both sides. If the numbers are not the same, draw the not-equal-to sign ≠ in the wreath.

CCSS.MATH.CONTENT.1.OA.D.7

Dapper Little Guy

Write the number of penguins hiding in the igloos to make the addition or subtraction equations true.

Oh Christmas Tree

In each row, circle the tree that is shorter by using the snowman as a guide for measuring.

CCSS.MATH.CONTENT.1.MD.A.1

Number the trees in order from smallest (1) to largest (3).

For Good Measure

Draw baubles inside the red boxes to measure the different objects. Be sure your baubles are the same size, they don't overlap, and there's no space in between them.

Example:

Santa hat = 5 baubles long

Now you try!

Snow globe = _____ baubles wide

Candle = _____ baubles tall

Candy cane = _____ baubles long

Tree = _____ baubles tall

Bells = _____ baubles across

Elf feet = _____ baubles wide

23

Most Wonderful Time

On this page, write the digital time in numbers beneath each clock. On the next page, draw the hands on the clocks to match the digital time.

:

:

:

:

:

:

:

:

:

2:00

3:00

5:30

7:00

11:00

4:30

1:30

12:00

8:00

Santa's Workshop

Use the graph to answer the questions below.

How many elves are in total?   + + _____

How many elves wrap gifts?   _____

How many elves decorate trees?   _____

How many elves make candy?   _____

How many more elves make candy than decorate trees?   — _____

How many more elves make candy than wrap gifts?   — _____

Organize the gifts beneath the tree into a graph by drawing the gifts in rows along the lines.

Draw which gift you have the most of: _____

Draw which gift you have the least of: _____

How many gifts do you have in total? _____

Merry and Bright

Every string of lights has exactly 120 bulbs. Some are already on the tree, but some still have to be hung. Count the bulbs that still need to be used to finish decorating the tree.

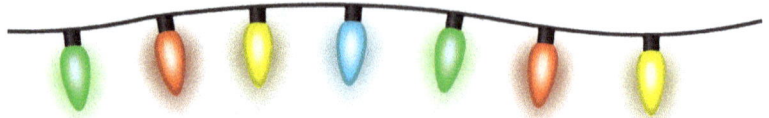

107 ___ ___ ___ ___ ___ ___

114 ___ ___ ___ ___ ___ 120

97 ___ ___ ___ ___ ___ ___

104 ___ ___ ___ ___ ___ ___

___ ___ ___ ___ ___ ___ ___ ___ ___ ___ 120

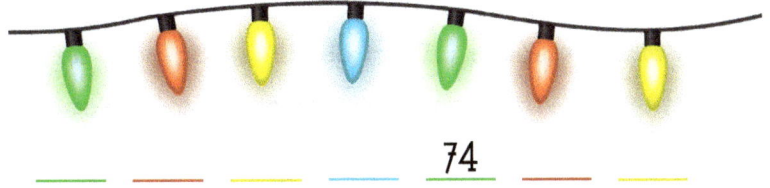

74

79

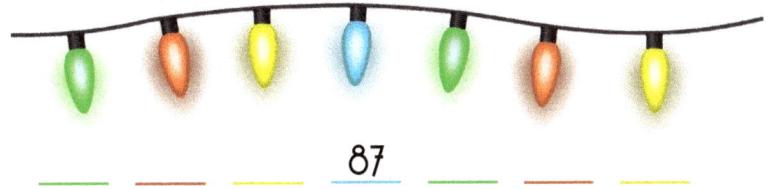

87

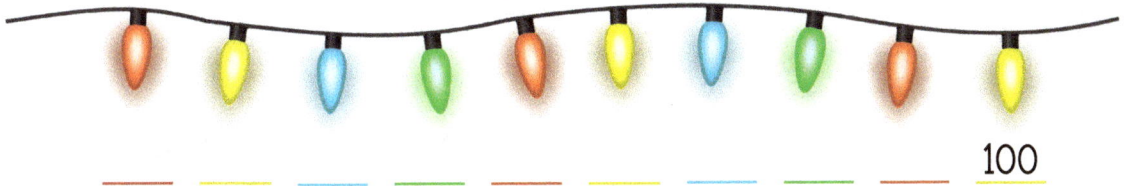

100

119

Clans of Gingerbread

In the white square within each candy box, write the missing number or draw the missing cookies. Every plate has exactly 10 cookies.

Tens   Ones

| 4 | 5 |

Tens   Ones

| | 3 |

Number & Operations in Base Ten

Tens	Ones
1	7

Tens	Ones
2	

Frosty Together Again

Build the snowmen back up out of the 10 different parts: hat, head, left eye, right eye, nose, mouth, scarf, belly, left arm, and right arm.

Tens the Season

The string of lights always surrounds 10 objects. Draw more objects in the candy cane box so they add up to the big red number.

 +  =13

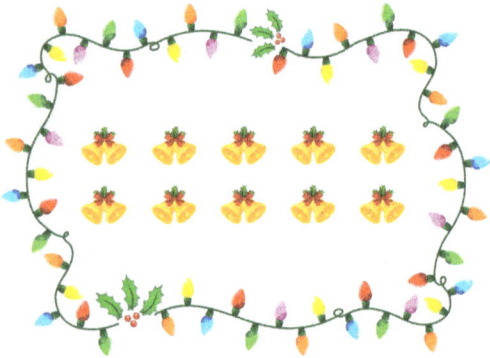 +  =14

 +  =11

CCSS.MATH.CONTENT.1.NBT.B.2.B

10 + $\square$ = 15

10 + $\square$ = 19

10 + $\square$ = 16

Gravy Trains

Each train car has 10 gifts. How many gifts are on each train?

_____ gifts

_____ gifts

_____ gifts

_____ gifts

_____ gifts

_____ gifts

Stocking Up or Down

Pretend the less than sign < and greather than sign > are beaks of hungry penguins. Inside each candy cane box, draw a "beak" facing the bigger pile of candy. Each stocking has 10 pieces of candy.

 32 ☐ 56

 42 ☐ 28

80 ☐ 60

13 ☐ 12

Boughs of Holly

Add the number of berries together. In the tens place, each bough of holly has 10 berries. Single berries are rolling around in the ones place.

Example:

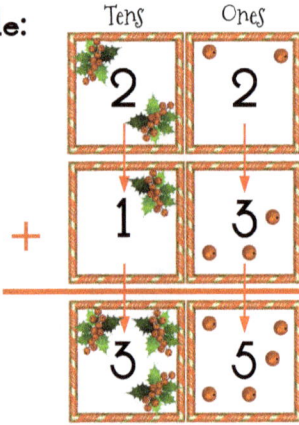

Number & Operations in Base Ten

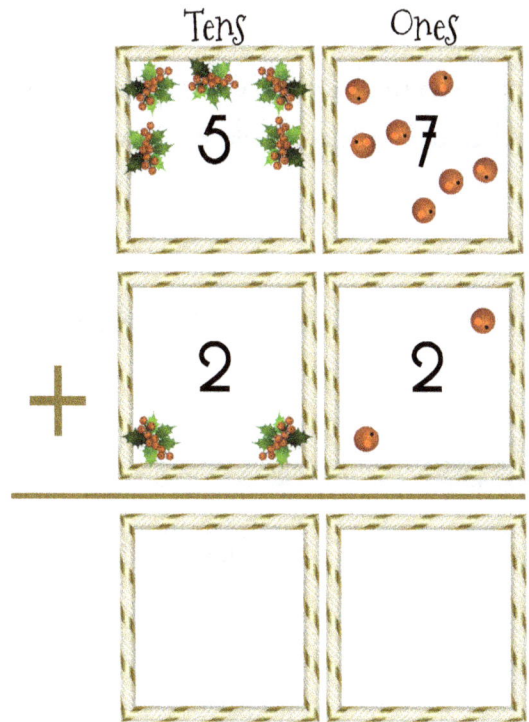

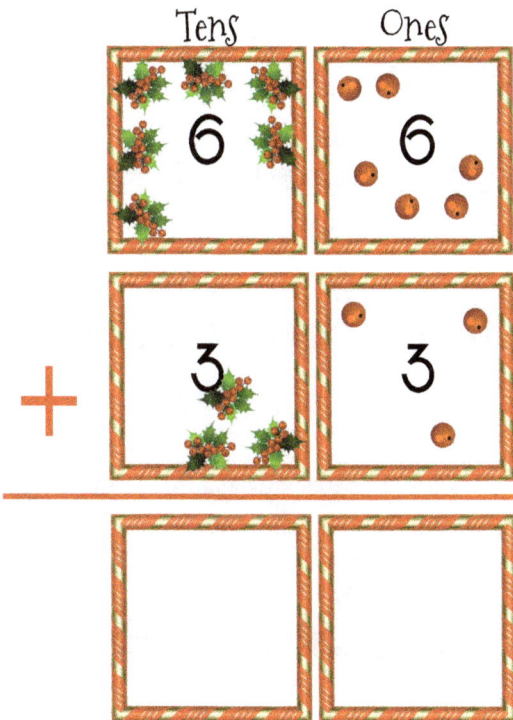

Candy Cane Lane

Each lane has 10 candy canes. With your imagination, find 10 more or 10 fewer total candy canes.

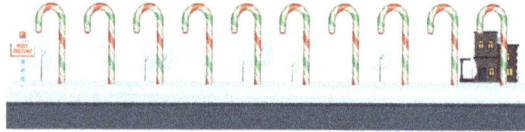

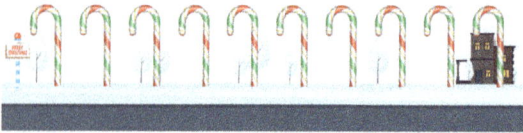

30 candy canes   10 more = _____

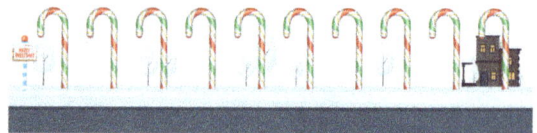

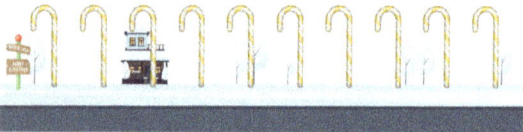

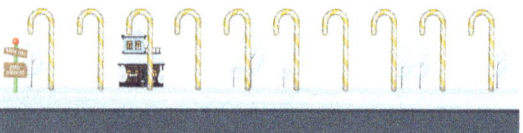

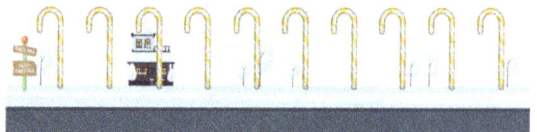

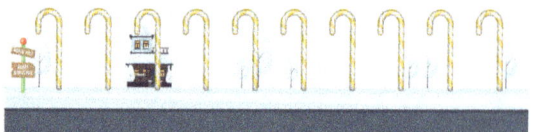

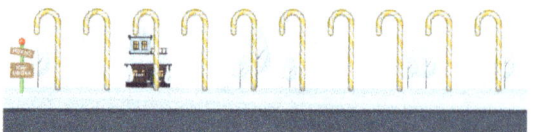

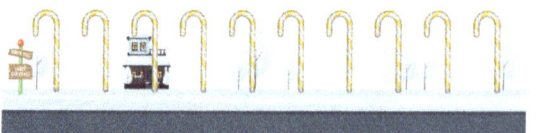

80 candy canes   10 fewer = _____

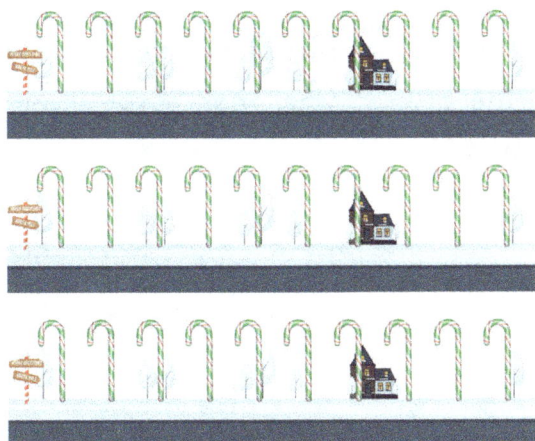

55 candy canes   10 more = _____

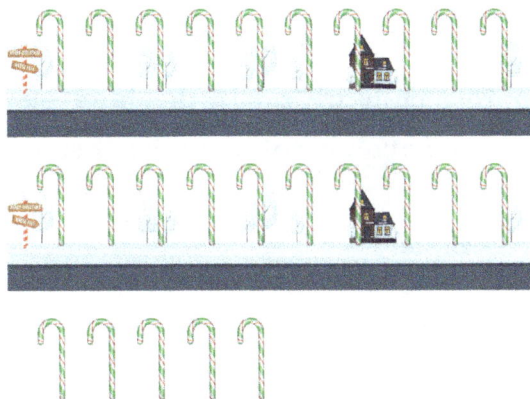

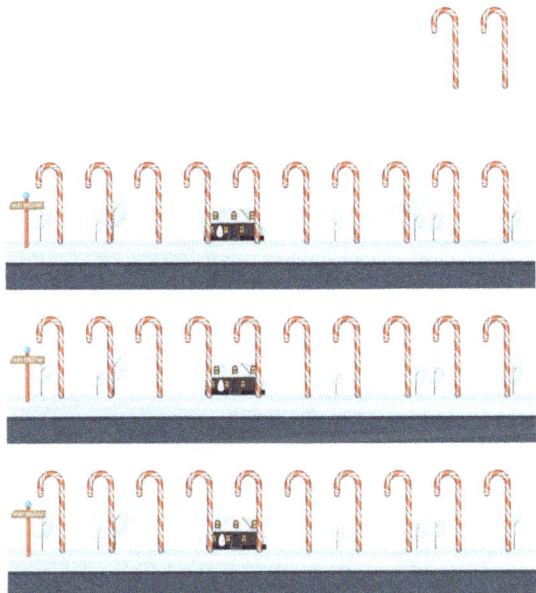

72 candy canes   10 fewer = _____

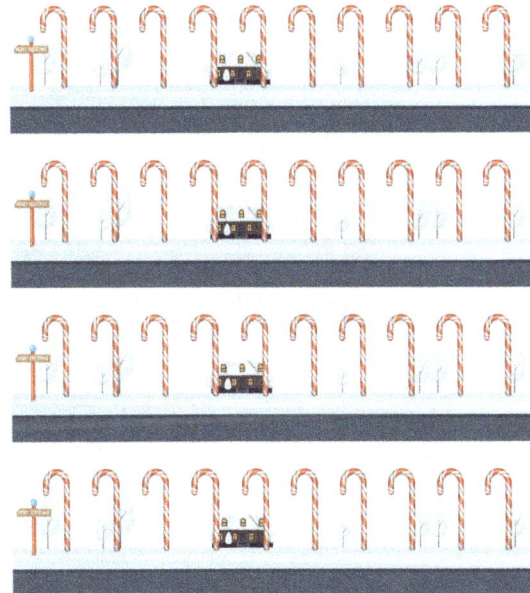

Red-Nosed Reindeer

Every reindeer has 10 points on his antlers. Subtract the number of points by working in the tens place value boxes. The ones place value boxes have no reindeer so they will stay at 0.

Example:

Tens	Ones
3	0
− 1	0
2	0

Tens	Ones
4	0
− 2	0

Tens	Ones
2	0
− 1	0

	Tens	Ones
	6	0
−	5	0

	Tens	Ones
	7	0
−	7	0

All Shapes and Sizes

Circle all circles, draw a triangle around all triangles, and draw a square around all squares. Cross out any shape that you feel is not a circle, triangle, or square.

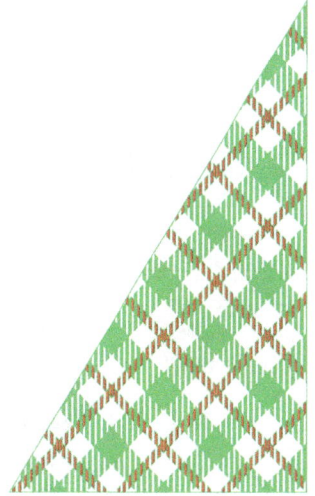

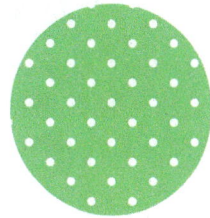

Shape Your Tree

Which of the single shapes make up the trees below? Trace the shape outlines and count how many shapes make up each tree.

circle

equilateral triangle

rectangle

trapezoid

square

right triangle

half circle

Total shapes = _____

Total shapes = _____

Use the shapes below to build your own tree!

How many total shapes does your tree have? _____

Season's Greetings

This bauble is cut in half. How many pieces are there?

_____ pieces

This bauble is cut in fourths. How many pieces are there?

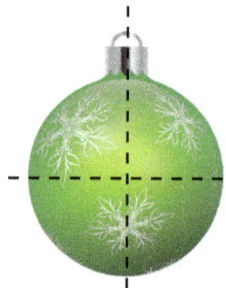

_____ pieces

Cut this bauble into fourths. How many pieces do you have?

_____ pieces

Cut this bauble into halves. How many pieces do you have?

_____ pieces

Circle the words <u>one half of</u> or <u>one fourth of</u> to make the sentence true.

is one half of / one fourth of a whole .

is one half of / one fourth of a whole .

This greeting card is cut in half.
How many pieces are there?

_____ pieces

This card is cut in quarters.
How many pieces are there?

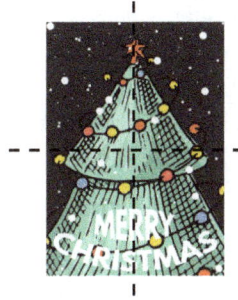

_____ pieces

Cut this greeting card into quarters.
How many pieces do you have?

_____ pieces

Cut this card into halves.
How many pieces do you have?

_____ pieces

Circle the words <u>one half of</u> or <u>one fourth of</u> to make the sentence true.

is one half of / one quarter of a whole .

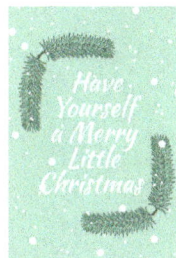

is one half of / one quarter of a whole .

OPERATIONS & ALGEBRAIC THINKING   COMMENTS

CCSS.MATH.CONTENT.1.OA.A.1 [p. 4]

Use addition and subtraction within 20 to solve word problems involving situations of adding to, taking from, putting together, taking apart, and comparing, with unknowns in all positions.

CCSS.MATH.CONTENT.1.OA.A.2 [p. 6]

Solve word problems that call for addition of three whole numbers whose sum is less than or equal to 20.

CCSS.MATH.CONTENT.1.OA.B.3 [p. 8]

Apply properties of operations as strategies to add and subtract.

CCSS.MATH.CONTENT.1.OA.B.4 [p. 10]

Understand subtraction as an unknown-addend problem.

CCSS.MATH.CONTENT.1.OA.C.5 [p. 12]

Relate counting to addition and subtraction.

CCSS.MATH.CONTENT.1.OA.C.6 [p. 14]

Add and subtract within 20, demonstrating fluency for addition and subtraction within 10. Use strategies such as counting on; making ten; decomposing a number leading to a ten; using the relationship between addition and subtraction; and creating equivalent but easier or known sums.

CCSS.MATH.CONTENT.1.OA.D.7 [p. 16]

Understand the meaning of the equal sign, and determine if equations involving addition and subtraction are true or false.

CCSS.MATH.CONTENT.1.OA.D.8 [p. 18]

Determine the unknown whole number in an addition or subtraction equation relating three whole numbers.

MEASUREMENT & DATA

COMMENTS

CCSS.MATH.CONTENT.1.MD.A.1 [p. 20]

Order three objects by length; compare the lengths of two objects indirectly by using a third object.

CCSS.MATH.CONTENT.1.MD.A.2 [p. 22]

Express the length of an object as a whole number of length units, by laying multiple copies of a shorter object (the length unit) end to end; understand that the length measurement of an object is the number of same-size length units that span it with no gaps or overlaps. Limit to contexts where the object being measured is spanned by a whole number of length units with no gaps or overlaps.

CCSS.MATH.CONTENT.1.MD.B.3 [p. 24]

Tell and write time in hours and half-hours using analog and digital clocks.

CCSS.MATH.CONTENT.1.MD.C.4 [p. 26]

Organize, represent, and interpret data with up to three categories; ask and answer questions about the total number of data points, how many in each category, and how many more or less are in one category than in another.

NUMBER & OPERATIONS IN BASE TEN

COMMENTS

CCSS.MATH.CONTENT.1.NBT.A.1 [p. 28]

Count to 120, starting at any number less than 120. In this range, read and write numerals and represent a number of objects with a written numeral.

CCSS.MATH.CONTENT.1.NBT.B.2 [p. 30]

Understand that the two digits of a two-digit number represent amounts of tens and ones. Understand the following as special cases:

CCSS.MATH.CONTENT.1.NBT.B.2.A [p. 32]

10 can be thought of as a bundle of ten ones — called a "ten."

CCSS.MATH.CONTENT.1.NBT.B.2.B [p. 34]

The numbers from 11 to 19 are composed of a ten and one, two, three, four, five, six, seven, eight, or nine ones.

CCSS.MATH.CONTENT.1.NBT.B.2.C [p. 36]

The numbers 10, 20, 30, 40, 50, 60, 70, 80, 90 refer to one, two, three, four, five, six, seven, eight, or nine tens (and 0 ones).

CCSS.MATH.CONTENT.1.NBT.B.3 [p. 38]

Compare two two-digit numbers based on meanings of the tens and ones digits, recording the results of comparisons with the symbols >, =, and <.

CCSS.MATH.CONTENT.1.NBT.C.4 [p. 40]

Add within 100, including adding a two-digit number and a one-digit number, and adding a two-digit number and a multiple of 10, using concrete models or drawings and strategies based on place value, properties of operations, and/or the relationship between addition and subtraction; relate the strategy to a written method and explain the reasoning used. Understand that in adding two-digit numbers, one adds tens and tens, ones and ones; and sometimes it is necessary to compose a ten.

CCSS.MATH.CONTENT.1.NBT.C.5 [p. 42]

Given a two-digit number, mentally find 10 more or 10 less than the number, without having to count; explain the reasoning used.

CCSS.MATH.CONTENT.1.NBT.C.6 [p. 44]

Subtract multiples of 10 in the range 10-90 from multiples of 10 in the range 10-90 (positive or zero differences), using concrete models or drawings and strategies based on place value, properties of operations, and/or the relationship between addition and subtraction; relate the strategy to a written method and explain the reasoning used.

GEOMETRY

CCSS.MATH.CONTENT.1.G.A.1 [p. 46]

Distinguish between defining attributes (e.g., triangles are closed and three-sided) versus non-defining attributes (e.g., color, orientation, overall size); build and draw shapes to possess defining attributes.

CCSS.MATH.CONTENT.1.G.A.2 [p. 48]

Compose two-dimensional shapes (rectangles, squares, trapezoids, triangles, half-circles, and quarter-circles) or three-dimensional shapes (cubes, right rectangular prisms, right circular cones, and right circular cylinders) to create a composite shape, and compose new shapes from the composite shape.

CCSS.MATH.CONTENT.1.G.A.3 [p. 50]

Partition circles and rectangles into two and four equal shares, describe the shares using the words halves, fourths, and quarters, and use the phrases half of, fourth of, and quarter of. Describe the whole as two of, or four of the shares. Understand for these examples that decomposing into more equal shares creates smaller shares.

www.ingramcontent.com/pod-product-compliance
Lightning Source LLC
Chambersburg PA
CBHW051803200326
41597CB00025B/4660